A large backcountry Yellowstone mudpot that is one of the best for producing bursts that resemble objects, both realistic and fanciful. This one reminds me of an aged alien with a cane, about 4 feet high. In some years, it acts like a spectacular mud geyser.

Look! What's that Lurking in the Mudpots

ISBN 978-1-944564-02-5
Printed in the U.S.A. by Sheridan Press

The sequel to this book, in similar format, will be titled More Things that Lurk in Mudpots.

Front Cover:

Although most photographs were taken in Yellowstone, the cover was not. It shows a small spitting mudpot located above a branch of the San Andreas Earthquake Fault adjacent to the Salton Sea in Southern California. It is driven by carbon dioxide percolating upwards from the shallow fault zone. The water is only warm to the touch, a pleasant alternative to all other locations used for this book.

I call this form of mudpot a splasher, since its water is quite fluid. It has a very shallow source of gas and fluid so is probably an offshoot from one of the more permanent mudpots and mud volcanoes nearby. The lifespan of a mudpot such as this is generally an hour or two.

It appears that a tree frog has climbed up to look through the window on the left. The total height of this mudpot burst is only about 4 inches!

A nearly circular panorama of the Salton Sea mudpot area as it is now. A geothermal plant has been constructed only a quarter mile away. Its steam emissions are visible at the far left. Surprisingly, many of the mudpots still remain, but nearly all the surface water is gone. Footprints endure, covering everything and the mudpots are an aged shell of their former appearance.

A fresh mudflow can be seen in the right foreground. In the blistering summer heat, these flows will whiten and crack apart within a day or so.

Look! What's that Lurking in the Mudpots?

Jack Hobart

An exploration into a world of imagination in which high speed photographs of mudpots reveal hidden shapes and patterns. Since mud is thicker than water, it form bizarre shapes that expand and ultimately break apart into pieces that are flung about by the hot steam within. The resulting shapes sometimes look like familiar forms, much like puffy clouds in the sky.

In addition to their twists and turns, the mud sometimes creates relatively quiet mud springs that contain darker impurities that form patterns like that shown above. These constantly change and reform, much like nature drawing upon a giant etch-a-sketch - creating a beautiful pattern that suddenly is erased by a bubble or upwelling in the mud that erases the pattern and starts another. Most are from back-country Yellowstone.

Most of the book consists of examples of bizarre mudpot images to challenge your imagination. I won't tell what I see in the picture captions, spoiling your fun.

Instead, the final section of the book contains picture thumbnails and interpretations from me and friends. Subsequent printings and updates to the electronic version may add feedback from readers. In the future, there may be a way to submit your feedback at ***www.mudpot.com***

Remember as you proceed, you never can tell what may be lurking in the pages to follow! Therefore, let your imagination run wild and join me in a fun adventure.

---

lurk . . . to be nearby, remaining hidden and lying in wait, avoiding observation by blending in with the surroundings. It suggests a sinister purpose, so one should beware of one's surroundings when in an unfamiliar place.

from the author

I'm an aerospace scientist, raised in Minnesota and living in California. I've been photographing mudpots in Yellowstone for over 30 years with many enjoyable memories shared with my sons, Craig and Chad, plus other friends including many members of GOSA, the Geyser Observation and Study Association. You can often recognize GOSA members by the group of people around them, soaking up their stores of interesting knowledge about the Park. Hope you run into them.

T. Scott Bryan edited the GOSA newsletter for the first decade and publishes The Geysers of Yellowstone, the go-to resource describing all geysers in Yellowstone plus descriptions of the World's geyser regions.

Tom and Genean Dunn took over publication of the Sput in 1996 and have tirelessly brought it up to the classy full color paper and electronic newsletter of today.

Rocco Paperiello and the late Marie Wolf compiled the most exhaustive descriptions of Yellowstone's thermal features including the 1,500 page "Lesser-Known Thermal Units of Yellowstone National Park". He is also an encyclopedic compendium of descriptive and historical information.

The Park Chief Geologist, the late Rick Hutchinson, was a supporter of the study of obscure park geothermal features including alerting me to new backcountry explosion craters. The Park Service and its rangers also provided invaluable assistance through the years, especially Sam Holbrook who led the Mudpot Special tours for many years.

The Goldbergs, Roz, Charlie, Dave and Mike also shared many of these features with me as did Mike and Cynthia Keller and their hawkeye daughters, Alex and Megan, who are particularly adept at discovering stuff that lurks.

The principal locations to see mudpots in Yellowstone are (1) the Mud Volcano area by the Yellowstone River, (2) the Lower Geyser Basin boardwalk and (3) the Artist Paintpots on the road to Norris. Ranger-led "Mudpot Special" tours were formerly taken to the many mudpots and mud volcanoes of Pocket Basin just northwest of Lower Geyser Basin.

The chapter Where are Mudpots? covers mudpots in other areas. Ron Keam in New Zealand and Andrey Leonov in Russia have shared their vast knowledge of thermal features in those countries.

One thought that I didn't have room to put at the end of this book. I hope that the Park Service can institute one message to all who visit the National Parks and every other place one visits:

"Wherever you go, please bring back one more piece of trash than you leave behind."

Please help me share this thought and set an example for all to help clean up America. Thanks !

Its time to explore the fascinating world of mudpots. There is an amazing variety of forms and patterns within them, even if you don't want to find objects that seem to lurk within them - lets go explore their sheer variety and beauty.

Good Hunting!!

Jack Hobart

A mud-splattered author after taking pictures at the mudpot that provided the most pictures for this book. A burst is occurring that obviously is much smaller than many others

contents

A tall 5 foot burst whose bubble is in the process of breading up. It was forced to this height by what is actually a rare mud geyser eruption column. This picture illustrates that a mudpot pattern is not just the mud itself, but can also be its shadow.

What are mudpots?

Mudpots are moderately acidic mud pools or vents in soft, clay-like soils that churn, bubble, erupt or explode intermittently as a result of gases rising toward the surface. The gas is superheated steam in volcanic areas such as Yellowstone. However it is warm carbon dioxide in the Salton Sea mudpots. These are located in an earthquake zone with some volcanic activity nearby as well.

If eruption sequences occur in distinct and somewhat predictable intervals, the mudpot is called a mud geyser. They behave like water geysers, however the thick mud produces smaller bursts. They are quite rare.

When ground water is not acidic, it deposits geyserite, which builds up the more solid structure typical of geyser basins. The two can coexist as well.

The mud forms different structures, primarily influenced by viscosity of the fluid. Viscosity is a measure of fluid thickness or resistance to flow. Thick mud can store up more energy. Therefore it can produce explosive bursts.

When the mud is fluid, swirling patterns can form on its surface as in the picture on the previous page. These consist of impurities such as black carbon from decomposed plants

When bubbling from below is more vigorous, splasher mudpots form, such as shown on Pages 9 and 43-47.

A normal mudpot bubble - about a foot across. It is starting to break apart, leaving some pieces hanging in place. Salton Sea, California

Some mudpots form by explosive breakouts, others by simply erupting through the soil followed by vigorous bubbling and churning action which widens the opening, forming a caldera looking much like its volcanic counterpart.

Mud is also thrown beyond the rim, raising it to form the mud counterpart to the ones formed by volcanic collapse.

As you can see, this is an exceptionally active one.

Any mudpot that bursts upwards can occasionally produce thin columns of mud.

A mud geyser threw columns to a height of 40 feet in the crater that produced many of the best bursts shown in this book. It only played as a geyser for a few years. Its spectacular shows are the subject of a future book, Pocket Mud Geyser. A short, wide column from those eruptions is shown on Page 6.

One of the complex splasher mudpot pools that creates wide patterns. This one is about a foot tall. Its mud is quite fluid. A black film can be seen in the background where the pool is quiet.

When a wet mudpot like that at the top of the page is only gently bubbling up, eddies or swirls are created as the mud is pushed outward into the thinner muddy water. This is an unusually nice and varied group. Close-up video of these is especially fascinating.

These two pages show the variability of patterns in the gentle bubbling that can occur in a quiet pool. Most of these pairs are close apart in time. For example, in the bottom pair, the large rising bubble in the left image is squashing the circular banding on its left into a face in the right hand image.

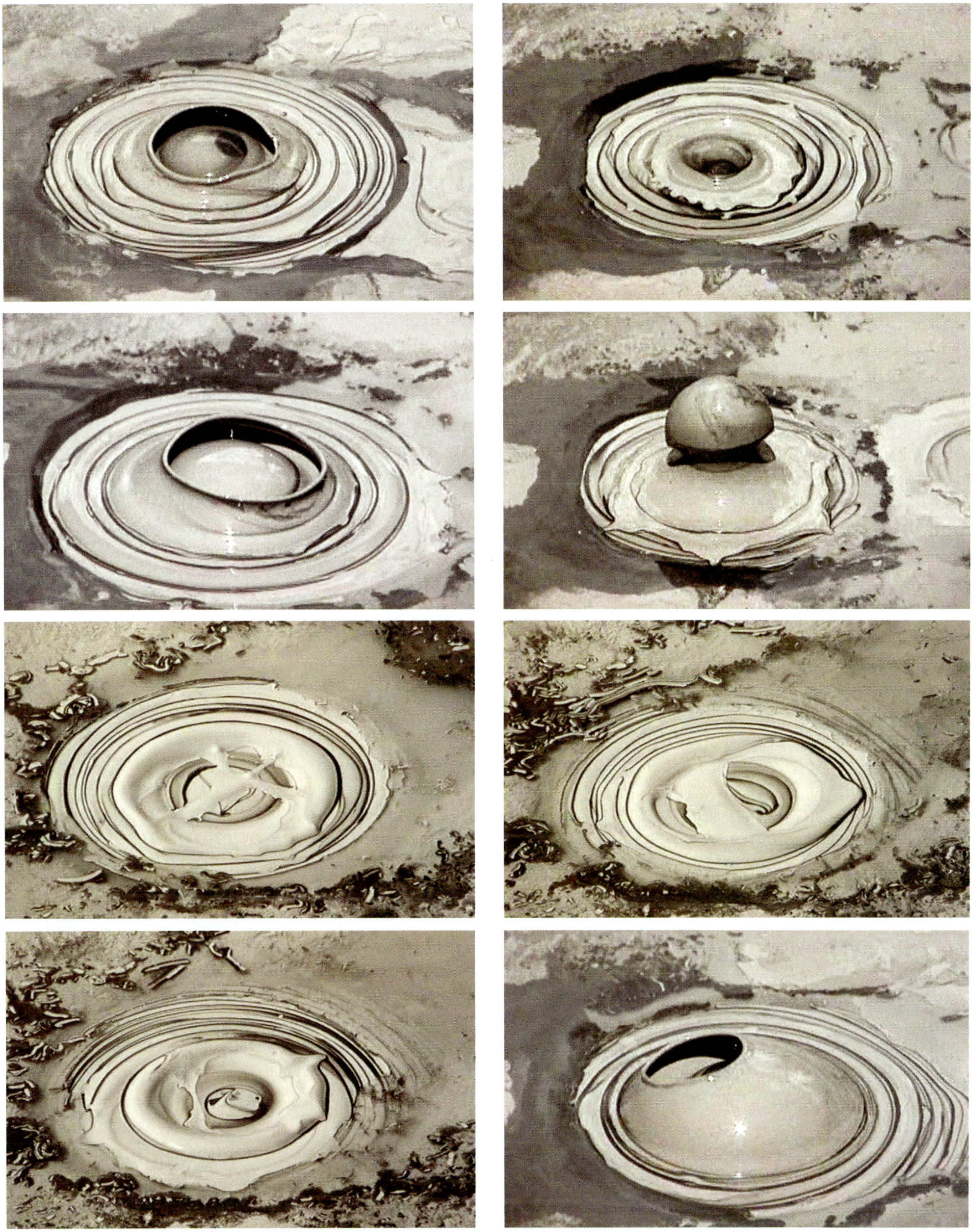

This small quiet pool in the remote backcountry has about 50 separate centers of quiet bubbling activity at any one time. I call it Kaleidoscope Mud Pool, because each time there is a ripple or bubbling up, the patterns change in remarkable ways, much like they do in a kaleidoscope.

This dry mudpot basin is composed of mud volcanoes, with a nice example in the foreground. A fresh mud flow has come down. The surrounding area is pock-marked due to rain falling on the mud before it had dried out.

The most dramatic mudpots are explosion craters. One breaks out every few years in Yellowstone. In 1987, two 70 foot craters broke out in a forest of lodgepole pine trees. Initially, the mud was dark black, a nice contrast to the white steam usually filling the crater and making photography quite difficult.

My son Craig looks at a violent mudpot burst from one of the larger Yellowstone mudpots. This is the color of most mudpots. The explosion crater on the previous page initially had black mud, however it turned to this color within a year or two.

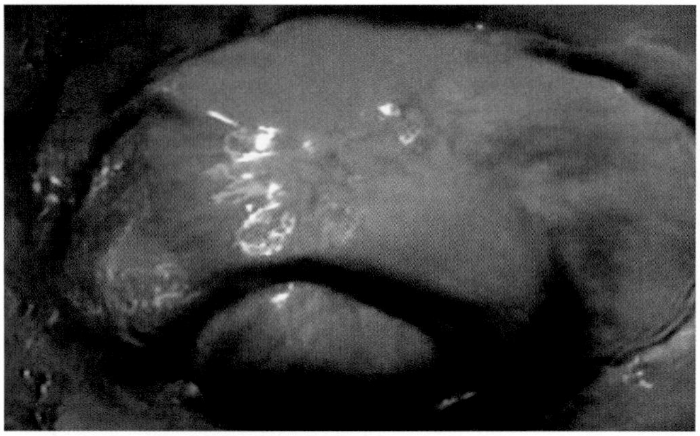
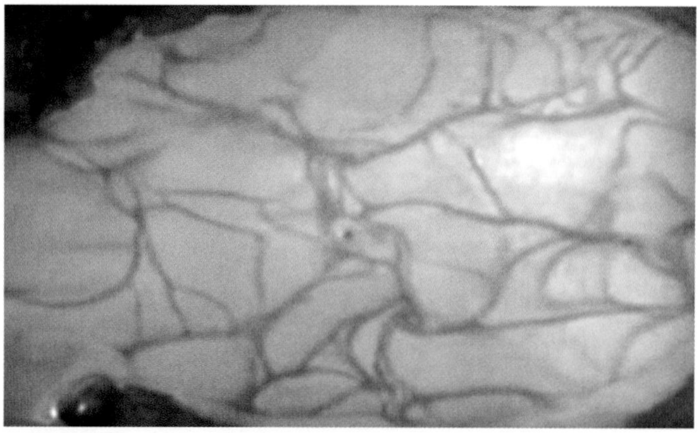
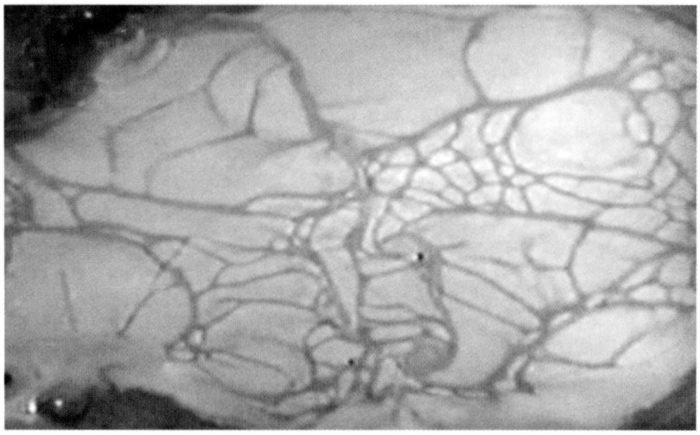
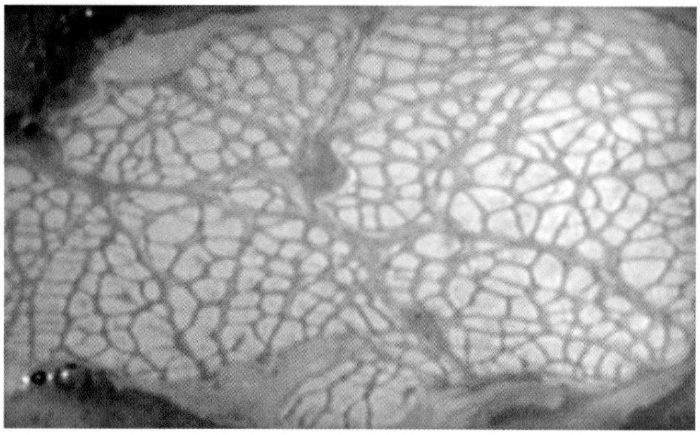

This is one of the most bizarre mud pools I have ever run across. It was a small plain pool, covered with scum. I happened to notice that bubbles were pushing the froth aside, revealing a plain, tan surface. Suddenly, lines began to appear, dividing the surface into more and more sections. On the video I was taking, I hear "My gosh, its cell division." Therefore, I named it Cell Division Pool.

It is behaving just like a culture of cells under a microscope, dividing in two, then four, etc. Next year it had stopped performing.

Much to my surprise, I discovered several similar mud pools within the Rotorua Golf Club in New Zealand, where sand traps are replaced by mud pots in the thermal area.

Here are three sequences from the video. The first picture above shows the bubble which pushes aside the froth and starts the cell division sequence. Below are two progressions of other cell division sequences..

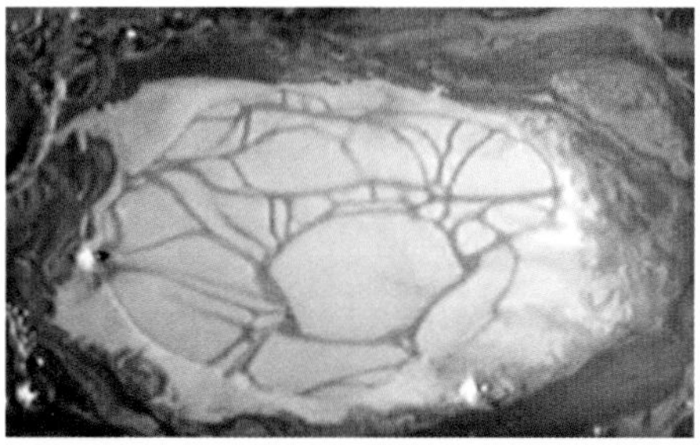
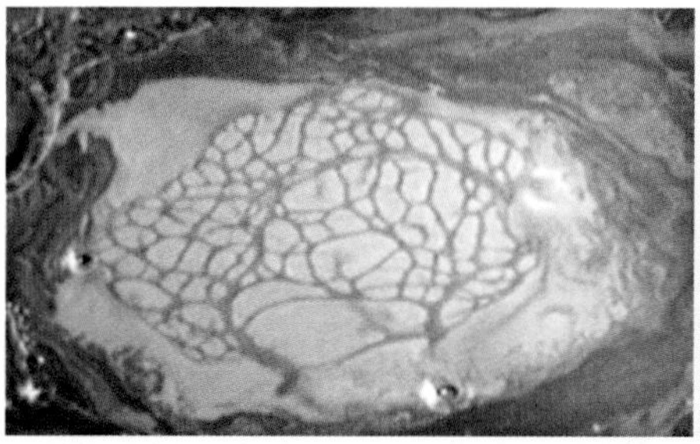
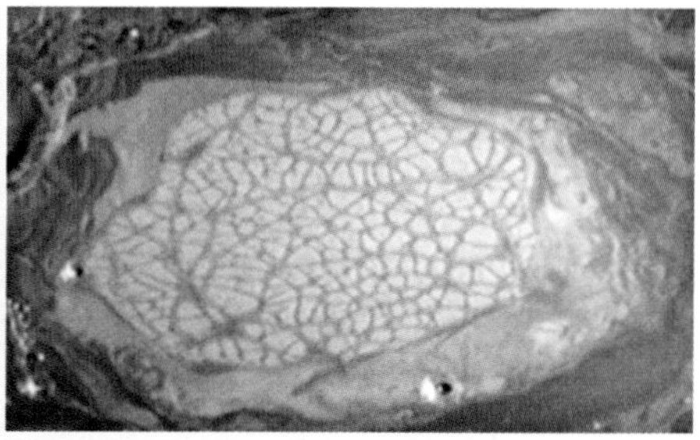
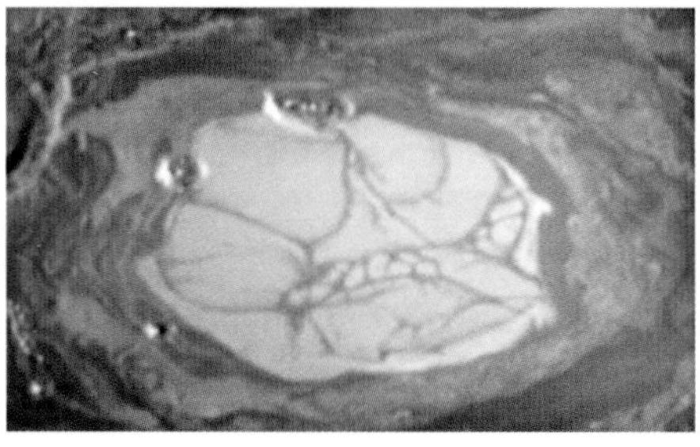
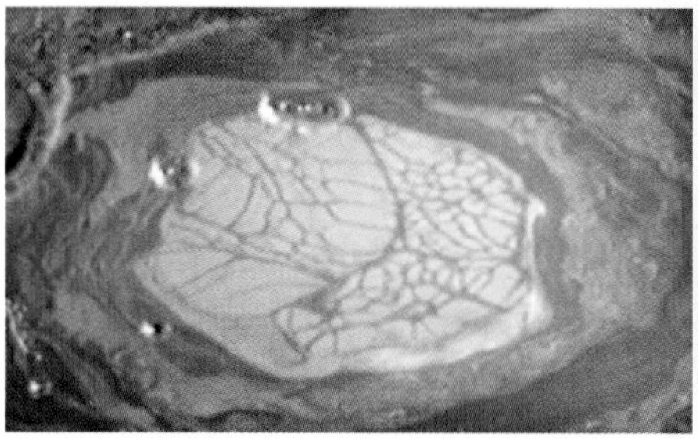
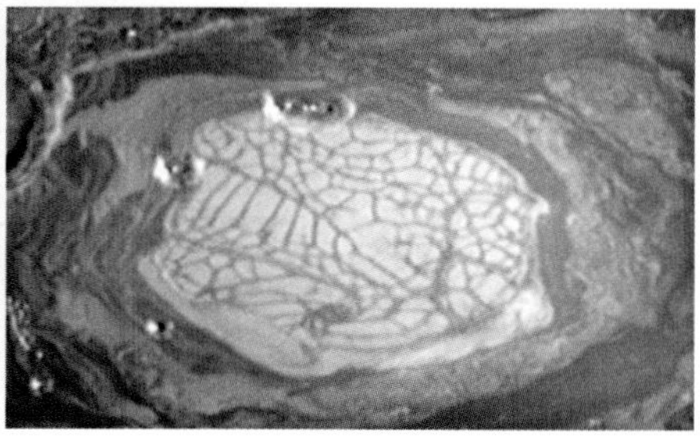

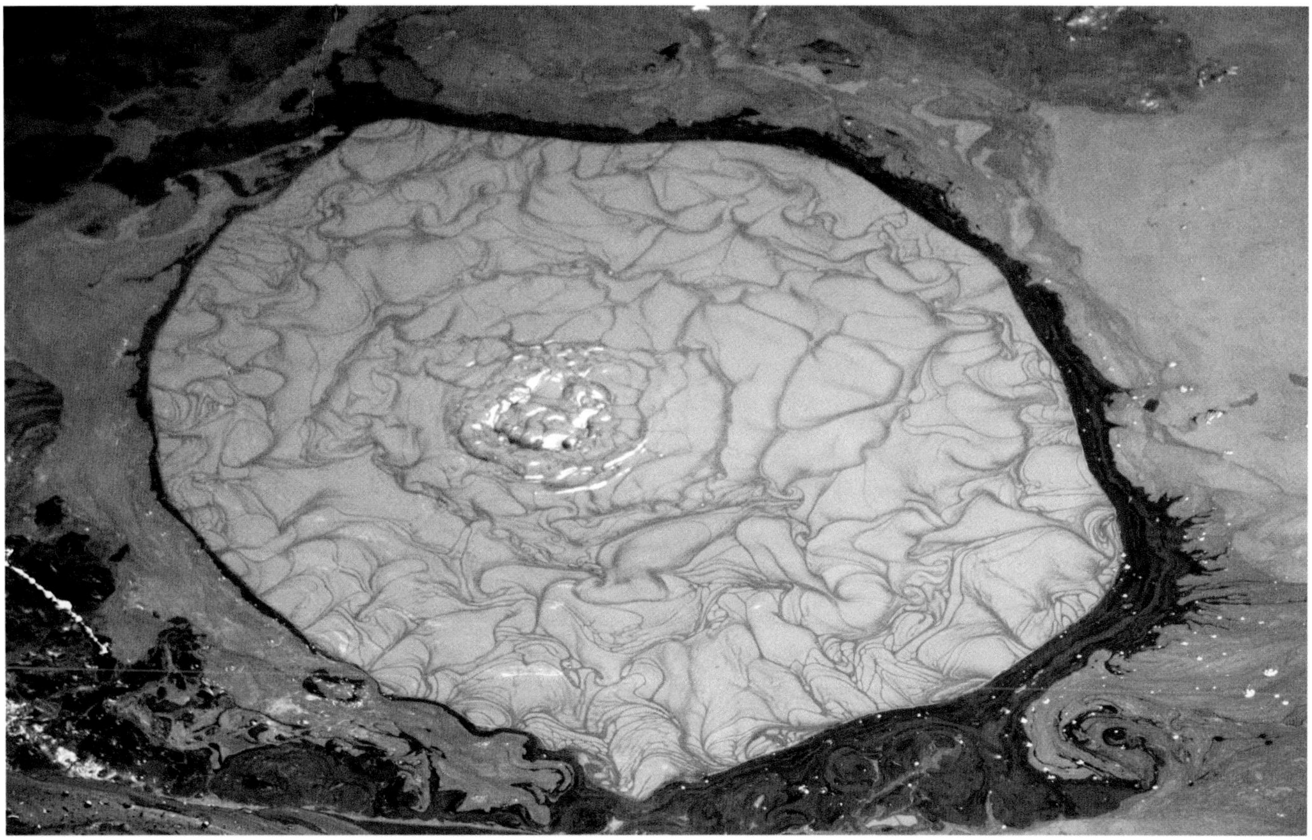

A cell division pool in a grove of trees in the golf course at Rotorua, New Zealand. The bottom image shows a higher degree of cell division. New bubbling is starting in the center of the mudpot.

Good friends Roz and the late Charlie Goldberg were studying one of the most unusual mudpots in Yellowstone. It is located trailside in the Washburn Hot Springs below Mt. Washburn.

The walls of these mudpots vibrate after nearby bursts. For this reason, I call these the rubber mudpots. When we first saw them, the sight of moving walls was so surprising and unexpected that we thought we were seeing things.

Their mud is black and foamy, like whipped pudding. It is an exception to the previous comment about typical mudpot colors.

The low wall the center of the picture is about 3 feet tall. Bubbling is occurring on each side of the wall, so it shifts back and forth continually with the changing levels of energy!

Imagine the difficulties of taking pictures of these features. Black on black is not a situation that comes up very often.

where are mudpots?

Mudpots are much more common than geysers throughout the World. They are also more tolerant of the ravishes exacted by geothermal drilling. When drilling occurred at the geyser fields in Nevada, the geysers immediately disappeared, however some mudpots remained.

The drilling adjacent to the Salton Sea mudpots in California eliminated nearly all surface water, however a fraction of the mudpots continue to endure.

Within Yellowstone, mudpots (including mud springs) are as widespread as geysers, but far less known.

I have been to only a few locations other than Yellowstone. Representative views from these places follow.

Below are several views at the Salton Sea from wetter times of old. After adjacent geothermal drilling, surface water is nearly gone and mudpot activity reduced.

New Zealand has the most interesting golf in the world. It took me 10 minutes to remove a golf ball from its crater in a "sand trap". Distracted by mud lakes, mudpots everywhere, cell division pools and 5 foot mud volcanoes in a lake, I managed to play only 5 of 18 holes of golf before dark, but got lots of photos and videos.

Iceland mudpots have much more steam in the freezing weather - even in June.

This mudpot photo appears to be taken at the same pool as used for the postage stamp. Its two bursts also mirror those on the stamp.

In July 1991, our group from GOSA was the first allowed to visit the Valley of Geysers in Kamchatka. There were no mudpots in the Valley, however we planned to trek to the Uzon Caldera, the primary location of mudpots. We called our July 4 celebration the day of horizontal snow. Three feet (1m) fell on the route, so we never got to see the mudpots. Thanks to Andrey Leonov, we have pictures of what the mud volcanoes and mudpots look like.

images with stuff lurking in them

The rest of the book consists of high speed images of these mud forms, mostly selected for the objects observed within the pictures. Some don't show obvious objects within their patterns, but you may discover hidden things lurking within them for yourself.

Captions will not disclose the identity of these objects so that you can guess for yourself.

An identifying number for each image is included. Number 01 is below. If you go to the chapter: Image Object Identification (Page 64), the objects others have found are disclosed. Each picture below has a corresponding numbered thumbnail in that chapter.

Hopefully you will soon be able to share your identifications of additional objects within the pictures. Persons who want to be identified would include their first names and ages. The website reserved for input submission is ***www.mudpot.com***

01. Starting off with something obvious. This should be the image for July in a mudpot calendar.

02. This remarkable burst is about 6 feet wide.

03. Taken from the same place as 02., 30 seconds later. I'll bet you don't come up with the same description as mine.

04. What I see here is bizarre. Maybe you can come up with something less fanciful.

05. There is something scary lurking in this mudpot.

06. This burst didn't seem like much until a very special subject was identified.

07. In contrast, everyone sees essentially the same thing in this burst..

08. This is a very puzzling object. After much study, several subjects jumped out. It is one of the more challenging bursts.

09. This could be a cute little object. Then again, rotating the picture could reveal another.

10. This figure is about 3 feet tall and is one of the most interesting ones. It gets many different responses.

11. This burst has multiple objects within it. Rotating the image helps. It is about 2 feet tall.

By now, you might have noticed similarities in some image backgrounds. This mudpot crater has produced many of the best objects. Another eruption site at its other end is usually more fluid, but makes many complex shapes as well as several of the best ones (# 02 and 03).

12. This larger and more complex burst looks quite weird. It's about 4 feet tall.

13. Kids do good with this one, especially if they have visited Disneyland. Its about 1.5 ft. across.

14. Another similar-sized object for kids. I'm not sure what this appears to be, but its distinct pattern must look like something special. .

15. The next few pages show Christmas-related images. This one is about 7 feet tall and is quite familiar to everyone.

16. This is a goofy Christmas image that requires much more imagination. Of course, there could be other objects within it as well.

17. This last Christmas object is the subject of my favorite Christmas card.

18. Not every subject has to be real. I let my imagination run wild for this one. Maybe you can find something more realistic.

19. Square openings shouldn't appear in fluid bubbles, but this sure looks like one. About 5 inches tall.

20. Another splasher mudpot with both a massive, irregular surface, plus bubbles being ejected upwards. About 1 foot wide.

21. This 5 foot tall object lives in one of the larger mudpots. I think that it is life-sized. My son Craig was looking at this one just like pictured on Page 13.

22. A strange little 2 foot object. Its hard to imagine how the smooth and textured surfaces were created and formed its textured column.

Sometimes video is being taken along with these pictures. Perhaps more information will be uncovered for a future update to this book.

23. A subject has actually been identified for this object. Hope you can come up an alternative. This is a large burst considering there is no spray at all - about 5 feet tall.

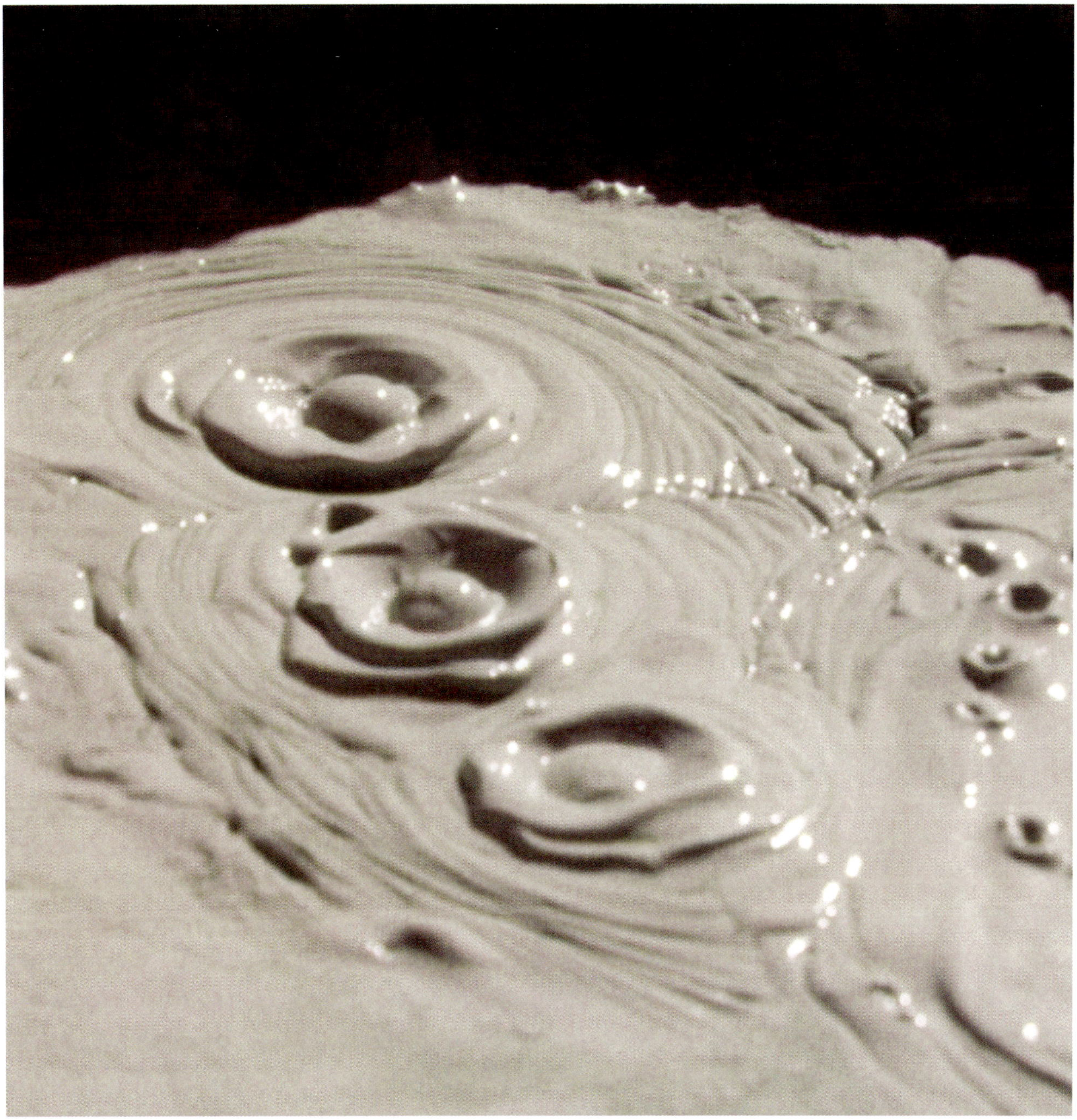

24. The next 5 images are all related. This one is quite obvious, however there are unusual variations to follow.

25. & 26. As promised, there are some twists to these images of the same subjects.

27. These 3 mudpot orifices are the same location as #25. A little change in the eruption bubbling as it comes to the surface can have a big change in the shape of the resulting objects.

28. A slight change in conditions turned these objects into something quite different.

29. Some interesting objects have been found it this burst, however you need to widen your thinking to find at least one of them. Can you think outside the box? I had a lot of problems with this one. One clue: sometimes pieces of mud seeming to hang in the air can help form hidden objects as well.

30. This one's a challenge. About a foot high and out of his element.

31. Another unusual shape. It's hard to understand how this shape could form. About 2 feet tall.

32. Wet splasher mudpots create complex shapes, often with all sorts of objects hidden within them. This one has some of my favorite objects. Similar bursts are shown on the following pages.

33. Another splasher with interesting shapes. Most of these come from a single obscure solitary backcountry pool. They are mostly just over a foot tall.

34. This one is low, wide and very complex. Could there be a story going on within it?

35. Closeup of another burst in the same mudpot. Its remarkable how conditions on a given day lead to similar shapes within different bursts.

36. Another identical shape from a third burst.

37. Yet another burst can create totally different results. The reason is that a widespread group of rising bubbles can interact in many different ways, producing a variety of outcomes.

38. Another splasher with some complex shapes.

39. This low one has some unusual projections above it. In this case, you get what you see.

40. This one's so complex and three-dimensional that there must be something lurking within it. There are also stringers and droplets all about.

41. Another splasher mudpot with activity similar to the previous one.

42. & 43. Two simple ones with single objects rising above them. They were included because I have a weird name for one of them.

44. The first of a series of images from the Salton Sea. It shows a typical setting for the mudpots, a foot wide bubbler on a 3 foot high mound that was constructed from a leveled field, probably 20 or 30 years ago.

This mudpot area is about 2 acres in size in a large field. When first seen in 1999, there were tire tracks passing over a 4 foot mound. Ten years later, that mound was nearly twice as tall.

This burst shows good examples of two quite different shapes. Both are shapes commonly seen in these interesting mudpots.

45. The same little splasher mudpot as shown on the cover. Its burst is about 6″ tall. It could bring terror to mind for anyone who sees what I see lurking in it.

46. One of the most rewarding images. It looks nice, but without any special patterns.

Now that you have sharpened your detective skills, the challenge is to find the most unusual object within this picture. I had to really open up my thinking for this one, but it was definitely worth the effort. I wish I had a recording of my exclamation when this object suddenly came to light.

47. This little spouter sits atop of one of the mounds. Its spitting has built up an 8″ cone extending outward from the top.

48. A small spitter mudpot somehow creates a complex grouping only about an inch high. It is the last in this series of pictures from the Salton Sea mudpots.

49. A complex low bubble in a mud pool. Thicker than the mud in the earlier splasher mudpot pictures shown from Yellowstone.

50. The complex start of a common burst shape for Yellowstone mudpots. It is nearly 2 feet high. Pieces are falling from the bubble as its column continues to push upwards.

51. A very strange-looking burst, about 2 feet tall.

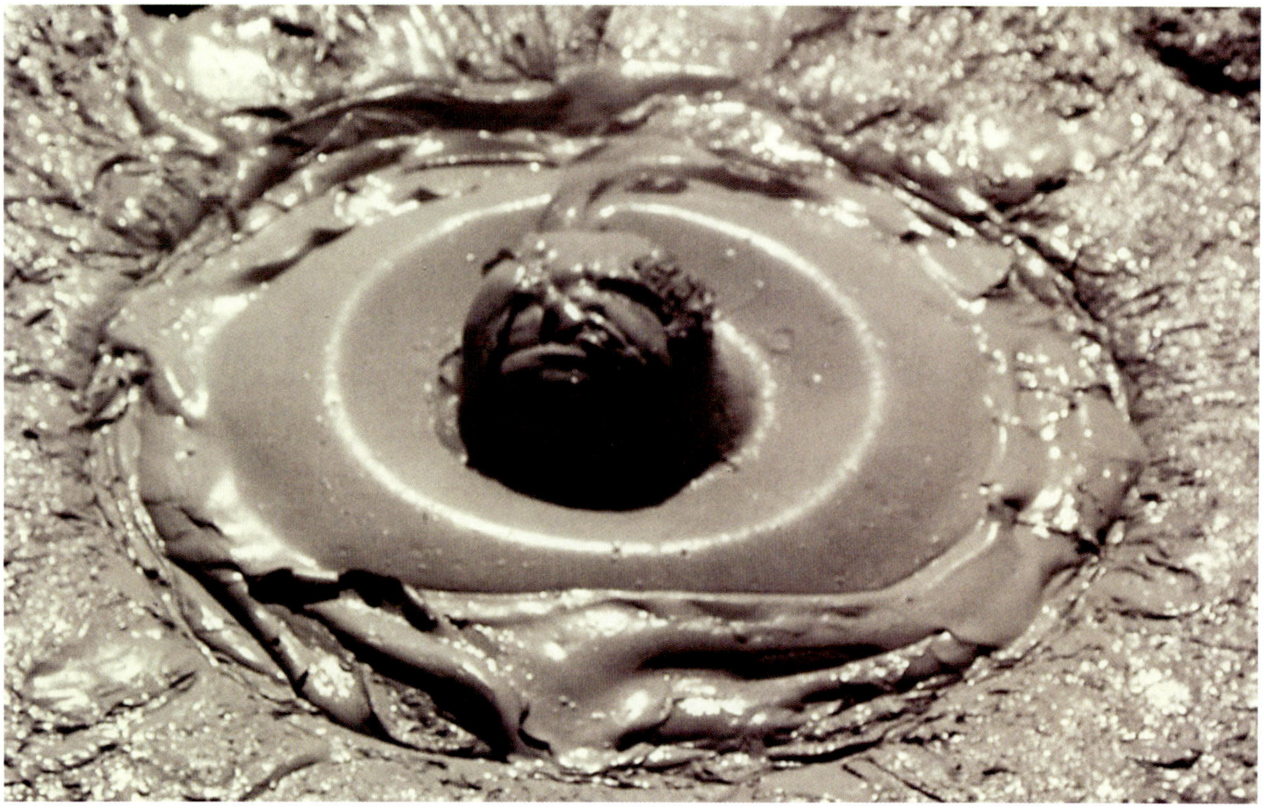

52. A thicker mudpot with folds left over from an earlier bubble burst. The outer ring is a little over a foot across.

53. A rare sequence of bursts. It appears that a short column followed by a piece of mud rose on the left. Next, an unusually-shaped burst in the center was followed by a column that is in the process of breaking into pieces. Such a pattern is a very rare occurrence. It is occurring in a mud geyser with periods of strong eruptions followed by long pauses - the subject of another book.

54. It appears that a tall bubble burst into large pieces. This may have been caused by the thick column erupting through the pieces. About 4 feet tall.

55. A burst sequence similar to that on the previous page. In this case, the earlier burst has exploded more violently and the subsequent burst rising through its center has an irregular shape. Ideal lighting conditions add to its beauty.

56. A burst from a crevice in one of the largest, deepest mudpot. It is about 4 feet tall. Its low, wide opening produces a low, throaty rumble that can be heard at great distances.

57. Another burst from the deep mudpot crevice. This one has an unusual pattern of smooth sections. It is also about 4 feet tall.

summary

Throughout this book, mudpot pictures have revealed characters lurking within the patterns. Hopefully, as you look around you, more objects will be revealed - not sinister or scary, but full of beauty and wonder.

These next few pages contain examples from Yellowstone illustrating that: wherever you look objects may be encountered that aren't what they appear to be, so look closely. They can appear in clouds, snowdrifts, floating debris in still water and even shadows. I'm sure you can discover far more by yourself.

A closeup view of a spring turned bubble-making geyser called Botryoidal Spring, located in the Lower Geyser Basin. The start of this eruption has produced a bird - a Thanksgiving turkey it seems.

This picture is from the book: **Unusual Geysers,** Hidden Bubble Patterns in Geysers of Yellowstone and Iceland. Park Service permission was given to collect data for this geyser from a close distance.

Some late afternoon scenes of mountains next to the sea shore with fog coming in?

Wait! Its just a muddy pool with sharp ridges looming above it, looking just like mountains in the dry desert southwest. They are actually only about 3 inches high

The fog rolling in consists of foam floating on the pool..

A coyote is chasing a grouse across a meadow with hot springs in it and mudpots nearby? Not really. Its just two tree roots in the grass!

"The Scream" painting by Edvard Munch. One of its four versions recently sold for over $119 million dollars.

"The Scream" - an instant in a mudpot burst that is eerily similar to this powerful masterpiece.

You have traveled through many types of images in this book, hopefully discovering all sorts of hidden characters and objects. Remember, if you encounter any scary monsters lurking about, you can always **SCREAM ! ! !**

image object identifications

The images in the section labeled "images with stuff lurking in them" starting with Page 20 have captions that don't identify any of the hidden objects within them. There are a few exceptions for some obvious "stuff". Each image has an identifying number which corresponds with the number associated with its thumbnail below.

Objects that have been recognized in the images are identified in a paragraph associated with each image.. Keep looking. I hope you find better objects or better names for them on your own.

01. An ice cream cone. The question is: Why does it appear to be growing out of a belly button?

02. "The Face". An incredibly detailed face with hair and a birthmark. Its texture looks like a Japanese kabuki mask. It is about 6 feet across!

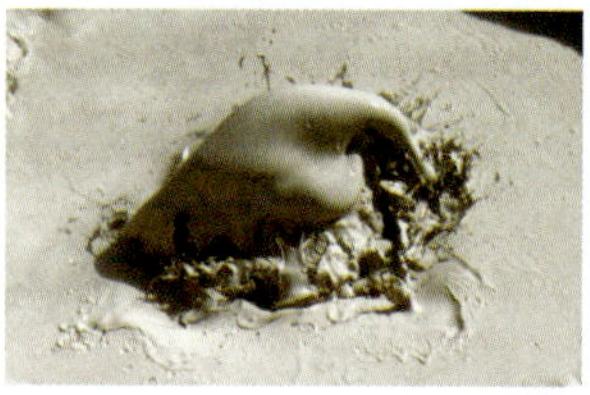

03. Anteater on a bed of lettuce. This appears to be a life-sized giant anteater, the ones from Patagonia. It is about 4 feet long. This picture was taken from the same location as 02. The Face, about 30s later.

04. A small column of mud is erupting on the left. An alien flying saucer has landed on the right and is investigating the burst using a periscope.

05. The Loch Ness Monster, "Nessie", has appeared in this mud pool. It has a small tail sticking up on the left. Perhaps it is old and tired, for its head could be propped up with a tripod.

06. Reclining Buddha. The seated image of Buddha is seen all over the World. Here, he seems to be leaning back from his seated position.

07. St. Christopher or another religious personage in robes.

08. This appears to be a face with large nose and cone-shaped hat. From the side, the top could be a finger pointing.

After several days, I had a revelation. I saw a wizard looking toward the right. He has a large, long nose and puffy chin, dressed in a robe. Several kids needed only a minute or two.

09. Appears to be a bare-bottomed baby/

10. People see many things in this image. A mushroom with its hand folded. A dog's head looking down to the left. A Darth Vader from Star Wars, also looking down to the left. A duck's head. The heads could be statues, mounted on the eruption column.

11. This is an interesting image. There are at least 3 heads, all quite different from one another. Rotate the picture to see them.

First, a round-headed Charlie Brown is at the bottom, looking to the left with his mouth open. Turning it upside-down, you see a long-nosed Wimpy, a cartoon character from the 1940's or so. Lastly, rotating it 90° clockwise, a sinister fish monster shows up with a scowl on its face.

Could there be more lurking inside as well?

12. More imagination needed for this one. Could be a shark jumping out of the water and beginning to swallow an eel or giant worm. But what's that on its head and back. Perhaps more ideas are needed.

13. "Hugging". It looks like two of the wooden characters from the Disney Small World ride hugging each other.

14. Similar style to #13, No idea of what it is. Could it be some sort of bathtub toy? Suggestions needed. At least, its an interesting and peculiar shape.

15. A fully-decorated Christmas tree. What is that on its top? Certainly no angel. Can you spot any unusual decorations? A magnifying glass may be needed.

16. Fat Santa. Could be some other character as well. What's he leaning back on?

17. This became my all-time favorite Christmas card. I call it: 'Frosty's Last Gasp "Merry Christmas" '.

A snowman is sinking in the mud pool, yelling out as his head sinks into the mud. One boy asked me, "Why is Frosty swimming like that?"

18. A Frankenstein mix of shapes. Looks like a seal or walrus with the head of a rooster.

19. Square openings should not be possible in a flat surface of a bubble. This appears to be an unusual vehicle with a window.

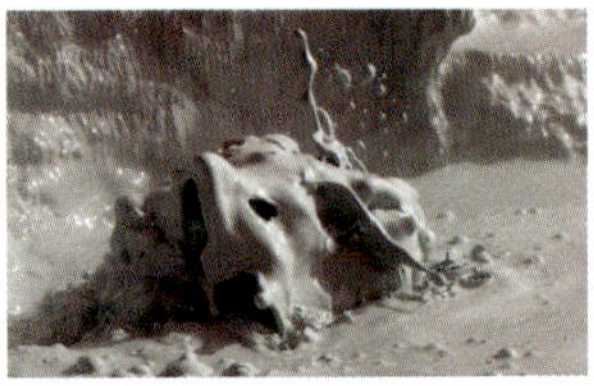

20. A complex shape with several heads looking upwards. This image is definitely up for suggestions.

21. This life-sized baby elephant is pictured in the bottom of one of the largest mudpots. It was the first mud object identified by me and my son Craig. It is about 4 or 5 feet tall.

22. Could be a small elephant girl with a dress on. She has little beady eyes and a very unusual trunk. About 3 feet tall.

23. "Elephant Debutante" This appears to be a demure or shy elephant all dressed up for a dance.

24. Three fried eggs. They start off the breakfast section of the book. These life-sized eggs result from smaller bubbles following larger ones. This is a very common sight in the mudpots.

25. A variation on the theme. A fried and a whole egg.

26. One of these eggs has a convenient carrying handle.

27. This is the same location as #25. It has 3 eruption sites. The large bubble on the right is just breaking. The small egg in the center formed earlier. On the left, some sort of character has just burst out of the mud and is calling out.

28. A slight change in the gentle bursts that create eggs forms a different object. I call this "two sombreros". Its interesting that both of these bursts differ from eggs by the same amount.

29. Time to open your mind to see these objects. Looks like a squinting Mr. Magoo with a skull cap on the top of this burst. It took a while to see it, but there is a face in a pumpkin in the center. This is a negative image with mud and shadowed openings trading places to form the image.

30. A fish with a large mouth. A bizarre image comes to mind: "Its time to anchor the fish" due to the anchor-like object sticking out of its tail.

31. One of my favorite objects. This unusual burst seems to be a head with large eye and streamlined helmet on a body made up of a large pedestal-shaped column. It would make a great metal sculpture.

32. This complex burst has several objects. In the center is a large crescent moon with a sleepy Mr. Magoo face. Topping it is a sculpted girl that could be a classic car hood ornament. On the right is a head with large nose, a mouth and closed eye. It is facing upwards. Writing this, I also see the head of a squirrel or ? with googly eyes, nose and mouth. It is in the bottom right. This is a great example of things lurking in the mudpot.

33. Another complex splasher. On the right is a bald man with moustache that could be a character out of The Wind in the Willows. At least three other heads are hidden elsewhere. One sideways with dark eyes, a bird-like head rising above it and a shrouded character in the background . . .

34. Even more complex. This low splasher introduces the monster with long head and no neck. Here he is about to swallow a pirate's head. The pirate has a patch on his right eye.

35. Another view of the same monster.

36. A third view of him. In this one he appears to have a gill.

37. Some sort of machine with claws. It is rising up out of this low burst.

38. An unusual pattern. Could this be a barbershop quartet of singers on the mound, looking up with mouths open?

39. An arrow and the letter "o" rise up above this low splasher.

40. This burst is rising higher, revealing some stringy objects. Several acrobats appear on the right. The left side could be a form of scaffold.

41. A series of flat surfaces that could be animals. Rising above them is a pair of horse-like figures. The rear one is a skeleton.
I'm stretching a lot for these descriptions.

42. The only thing that stands out in this simple burst is what I call "The Last Republican" from the lone elephant figure.

43. Another simple one with an irregular figure rising up.

44. The start of a series from the Salton Sea mudpots. A facemask fills the mudpot with a twisting column rising above it. These are the two common eruption forms at these mudpots.

45. "At the Dentist". A character with a strange single lock of hair (or a weird cap). He has his mouth wide open and a dentist drill in it. Ouch!

46. This simple mudpot column needs to be rotated clockwise to see the hidden and very subtle object.

The surprising find: a head is being shot out of a cannon. Stretching out, he has a pug nose, eyes wide open and is yelling at the top of his lungs. A truly remarkable object and probably the most hidden of all.

47. A head from behind, looking to the right. It extends outward from the top of one of the mounds throughout this area.

48. A dog leaving a doghouse. This is only an inch high.

49. An octopus is lurking in this pool.

50. A jellyfish. This object will become a mushroom when the pieces fall to the surface of the pool.

51. A bizarre object. I don't know what it is, but I call it "Swallowing His Stomach". The thing has a very big nose.

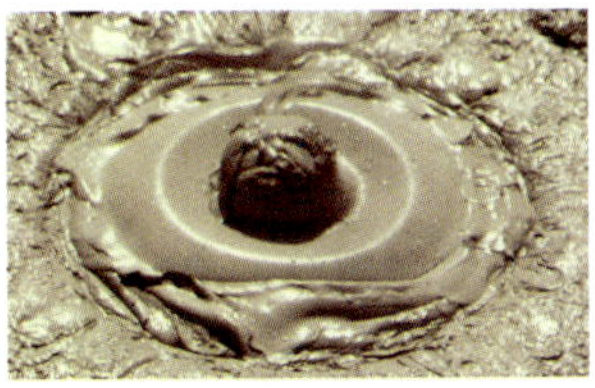

52. A platter with something in its center. Could be a baked apple . . . or something really gruesome. At Halloween, it takes on a whole new meaning.

53. An interesting combination of features. The largest object looks like a helmeted little guy with arms outstretched. Earlier in the eruption, when the pieces were connected, it could have looked like a tall guy with a sheathed hand carrying an upright sword.

54. Nothing comes to mind, but its a nice combination of pieces.

55. A pretty tree standing out boldly against the shady background.

56. With enough imagination, this could be some animal being attacked by another on the right. However, its head looks more like a parrot than a meat-eater .

57. Could be a dinosaur, turning its head to bite its backside like a dog does.

Several closeups of tall mudpot splashing bursts - actually less than two feet tall.

Their droplets form when filaments are projected above the surface, then contract due to surface tension pulling at their surfaces, causing them to break into droplets

The bottom one could be a climber scaling a tall, narrow mountain with a snowboard in his pack. But then, I could be wrong.

postscript

As you can tell by now, I am fascinated by the incredible variety of shapes and patterns produced by mud, whether in the form of bubbles, bursts, explosions or simply patterns etched in a pool surface by upwelling mud of a different color.

At the very least, these mudpots produce an interesting variety of shapes that are quite different from those resulting from water splashing.

As a physicist, I continually marvel at the effects of surface tension, the attractive force at the molecular level. It seems to magically transform or contort the mud surfaces into an endless variety of shapes.

I am so thankful that the high speed cameras of today enable one to capture instants of time that cannot be clearly seen with the naked eye.

While taking these pictures, hundreds of hours of video and high shutter speed video have also been taken. When paused or played back at slow speeds, the evolution of the bubble bursts can be seen, providing even more wonder at these spectacles. The addition of mudpot sounds, the pops, splashes, gurgles, bangs and even rhythmic drumming add another dimension to them. These sounds are a major reason that mudpots are so interesting to tourists at Yellowstone. Its planned eventually to share the sounds and some video of mudpots at the website: ***www.mudpot.com***.

In one sense, it is fortunate that these mudpot bursts are mostly shades of gray or tan. If they were in a kaleidoscope of color, one couldn't focus on the shapes and patterns that make these mudpots so special. I have converted images to false colors using the computer, however any more than a tweek to the colors rapidly turns to more than my brain can take in.

One regret is that this book couldn't have been done earlier so that it could have been given to Walt Disney for inspiration of something fun.

In closing, I'd like to share a small view of my favorite mudpot photograph with you. A tiny foot-wide splashing burst taken by lying down nearby at the only time of day that shadows appear on the edge of the pool in the background. It looks like the view of an alien world, so I call it: "Planet Landscape". After years of gazing at it, I just discovered several faces hiding in plain sight right in the center of the image!

Jack Hobart

Planet Landscape

books by the author

Mexican Opal Canteras
Spectacular Cabs of Fire
2015
softbound & electronic
88 pages 133 images

Unusual Geysers,
Hidden Bubble Patterns in Geysers of Yellowstone & Iceland
2016
softbound
136 pages 237 images

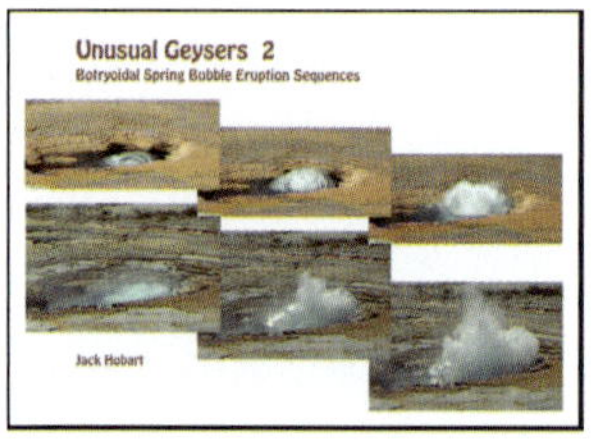

Unusual Geysers 2,
Botryoidal Spring Bubble Eruption Sequences

pdf e-book only
224 pages 240 images

Look! What's that Lurking in the Mudpots
"Through the Looking Glass"
2016
softbound
72 pages 117 images

More Things that Lurk in Mudpots
"Witch way did they go?"

softbound
80 pages 156 images

Pocket Mud Geyser,
The Eruption Years 1994-2004

softbound & electronic formats
96 pages 180 images

Skip an Atom Agates
The Most Unusual Lake Superior Agates
2016
softbound
96 pages 276 images

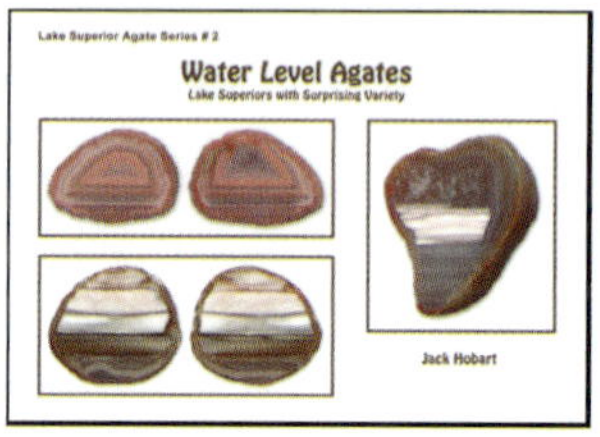

Water Level Agates
Lake Superiors with Surprising Variety

softbound
112 pages 217 images

Lake Superior Copper Agates
Introduction to the World's Most Spectacular Agates
2016
softbound
88 pages 175 images

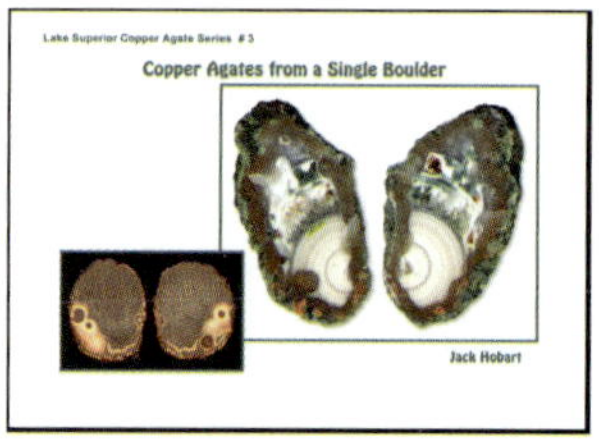

Copper Agates from a Single Boulder

pdf e-book only
88 pages 146 images

Tiny Lake Superior Agates
Treasures Anyone can Find
2016
softbound
64 pages 141 images

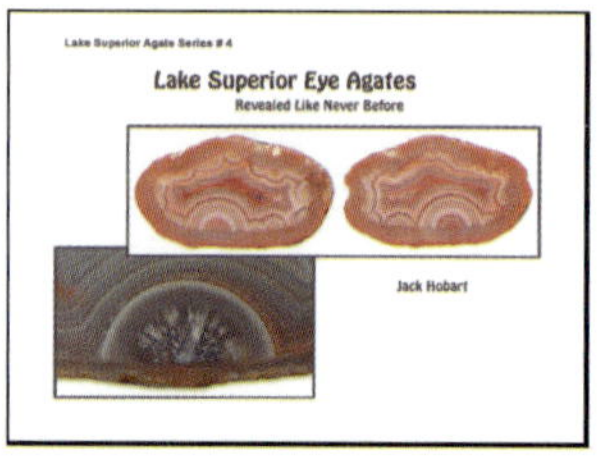

Lake Superior Eye Agates
Revealed Like Never Before

softbound
116 pages 179 images